ASTONISHING ANIMALS

by Kristin Marciniak

www.12StoryLibrary.com

Copyright © 2018 by 12-Story Library, Mankato, MN 56003. All rights reserved. No part of this book may be reproduced or utilized in any form or by any means without written permission from the publisher.

12-Story Library is an imprint of Bookstaves and Press Room Editions

Produced for 12-Story Library by Red Line Editorial

Photographs ©: TheSP4N1SH/iStockphoto, cover, 1; Norbert Wu/Minden Pictures/Newscom, 4, 5; Alias 0591 CC2.0, 6; April CC2.0, 7, 28; francesco de marco/Shutterstock Images, 8; kajornyot/iStockphoto, 9, 29; Stephen Dalton/NHPA/Photoshot/Newscom, 10; Derek Keats CC2.0, 12; Miki Simankevicius/Shutterstock Images, 13; Gerard Lacz/VWPics/Newscom, 14, 15; Amy Goldstein CC2.0, 16; Jan Mastnik/Shutterstock Images, 17; Ted Treska/USFWS, 18; Pedre/iStockphoto, 19; Michelle Valberg/Glow Images, 20; Christian Darkin/Science Source, 21; Phillip W. Kirkland/Shutterstock Images, 22; Tony Alter CC2.0, 23; 2630ben/Shutterstock Images, 24; Foto Mous/Shutterstock Images, 25; Becky A. Dayhuff/NOAA, 26; Sphinx Wang/Shutterstock Images, 27

Content Consultant: Julia K. Whittington, DVM, Clinical Associate Professor, Department of Veterinary Clinical Medicine, University of Illinois

Library of Congress Cataloging-in-Publication Data
Names: Marciniak, Kristin, author.
Title: Astonishing animals / by Kristin Marciniak.
Description: Mankato, MN : 12 Story Library, [2017] | Series: Unbelievable | Audience: Grades 4 to 6. | Includes bibliographical references and index.
Identifiers: LCCN 2016046436 (print) | LCCN 2016054117 (ebook) | ISBN 9781632354174 (hardcover : alk. paper) | ISBN 9781632354877 (pbk. : alk. paper) | ISBN 9781621435396 (hosted e-book)
Subjects: LCSH: Animal behavior--Juvenile literature.
Classification: LCC QL751.5 .M334 2017 (print) | LCC QL751.5 (ebook) | DDC 591.5--dc23
LC record available at https://lccn.loc.gov/2016046436

Printed in the United States of America
022017

Access free, up-to-date content on this topic plus a full digital version of this book. Scan the QR code on page 31 or use your school's login at 12StoryLibrary.com.

Table of Contents

Anglerfish Lure Prey with Light 4

Atlas Moths Disguise Themselves as Snakes 6

Cuckoos Get Others to Raise Their Young 8

Green Basilisks Sprint on the Water's Surface 10

Honeyguides Help Humans Hunt for Hives 12

Hooded Seals Blow Big Balloons 14

Hungry Dung Beetles Dine on Manure 16

Lampreys' Jawless Mouths Latch onto Prey 18

Narwhal Tusks Have Super Senses 20

Opossums Stage Dramatic Death Scenes 22

Pangolins Protect Themselves with Armor 24

Sneaky Octopuses Outsmart Predators 26

Fact Sheet 28

Glossary 30

For More Information 31

Index 32

About the Author 32

1

Anglerfish Lure Prey with Light

Female anglerfish lure prey with their photophores.

A fierce predator lurks in the darkest parts of the ocean. A female deep-sea anglerfish lies in wait for unsuspecting prey. Long, jagged teeth angle inward toward its gaping mouth. But its terrifying appearance is not what makes the anglerfish astonishing.

Female anglerfish hunt other fish. They have long spines near the top of their heads. The spines act as fishing rods. A fleshy nub called a photophore serves as the bait. The photophore gives off a blue-green light. The light is created by a process called bioluminescence. It happens when a chemical in the anglerfish's body mixes with oxygen. This process is also what makes fireflies glow.

The female anglerfish uses the light from its photophore as a lure. The photophore catches the attention of fish and marine

1.9
Distance, in miles (3,000 m), below the ocean's surface a deep-sea anglerfish can survive.

- Female deep-sea anglerfish use spines as fishing rods.
- A photophore on the spine's tip emits a blue-green light to attract prey.
- Male anglerfish do not hunt, but depend on females for food.

invertebrates. When her prey swims by, the anglerfish snaps her jaws shut. Then, she swallows her prey whole.

Female anglerfish are deep-sea predators. But male anglerfish do not hunt for prey. They rely on female anglerfish for food. A male anglerfish has sharp teeth. He latches onto a female's body. His lips fuse to her body. This connects their blood vessels. Six male anglerfish may attach to a single female. Male or female, anglerfish are astonishing deep-sea fish.

An anglerfish's jaw allows her to catch and swallow her prey whole.

2

Atlas Moths Disguise Themselves as Snakes

The atlas moth of Southeast Asia is the largest moth in the world. It has a 12-inch (30-cm) wingspan. That is longer than a sheet of notebook paper. These brown, hooked wings are intimidating. They look like twin cobra heads.

The atlas moth is a mimic. It has evolved to look like another animal. The pattern on its wings tricks hungry birds.

The tips of an atlas moth's wings look like snake heads.

1
Number of weeks the adult atlas moth lives.

- The atlas moth of Southeast Asia is the largest moth in the world.
- Its wings look like the heads of two cobras.
- Mimicry prevents predators from eating adult moths before they can mate and lay eggs.

The birds think the moth is a poisonous snake. When threatened, the moth drops to the ground. It slowly moves its wings, making its "cobra heads" look alive. Real cobras eat birds. Birds avoid anything that looks like one.

Atlas moths do not live long. They are unable to eat. Instead, they live on fat stored during their time as caterpillars. The moths have only enough time to mate and lay eggs before they starve.

MORE MIMICS

Mimics are found throughout the animal kingdom. But butterflies and moths are some of the most recognizable. The giant owl butterfly has black and yellow spots. The spots make it look like an owl's eyes. The eyed hawk moth is colored to look like a pair of fox's eyes. Mimicry is not limited to adult insects. Several species of caterpillars are colored to look like snakes to scare potential predators.

Giant owl butterflies mimic an owl's eyes.

Cuckoos Get Others to Raise Their Young

The cuckoo is a small bird. But it is one of the biggest bullies in the bird world. Cuckoos are brood parasites. They do not raise their own chicks. Instead, a female cuckoo lays her eggs in the nest of another bird. This host bird cares for the eggs. It raises the cuckoo chicks.

Cuckoos have adapted to give their chicks the best chance at survival. Some cuckoo eggs are colored to look like the eggs of the host bird. This tricks host birds into thinking all of the eggs in the nest are theirs. Other cuckoos toss the host's eggs out of the nest or eat them. This decreases competition from other chicks.

Cuckoo chicks are sometimes much larger than the host birds that care for them.

> A common tailorbird feeds a much larger cuckoo chick.

Cuckoo chicks know how to survive before their eyes are even open. They muster every ounce of strength in their weak, featherless bodies. Then, they push the host's eggs or chicks out of the nest.

Why do host birds raise chicks that are not their own? In many cases, cuckoo chicks share features with the host bird's chicks. They might have similar calls for food. The host bird cannot ignore its instinct to feed a hungry chick.

13
Maximum number of eggs left in one nest by a great spotted cuckoo.

- Cuckoos rely on other bird species to raise their young.
- Mother cuckoos may destroy the host's eggs before laying her own.
- Newborn cuckoos push other eggs out of the nest to increase their own chances of survival.

4

Green Basilisks Sprint on the Water's Surface

The rain forests of Central America can be dangerous places for lizards. The green basilisk's grassy hue helps it hide from predators. But sometimes a predator gets too close. When this happens, the green basilisk jumps into the nearest stream. But it does not swim below the water's surface. Instead, it runs on top of it.

The green basilisk appears to walk on water.

CAMOUFLAGE

Camouflage is a combination of colors and patterns that disguises animals. It helps them hide from predators or prey. The green basilisk is the same color as many leaves in the rain forest. This camouflage helps the lizard blend in with its surroundings. Other animals have camouflage that makes them look like something else. Young swallowtail caterpillars look like bird droppings. This discourages hungry birds from eating them.

Basilisks are bipedal in the water. This means they run on two legs, as humans do. Their strong hind legs whirl like windmill blades. The green basilisk is very lightweight. It moves fast, approximately 5 feet per second (1.5 m/s).

Each slap of the basilisk's long toes against the water creates tiny air pockets. These air pockets keep the lizard from sinking below the surface. The lizard can sprint for more than 15 feet (4.6 m). Then, gravity kicks in. But sinking is not a problem for basilisks. They are also great swimmers. Adults can stay underwater for up to 30 minutes.

65

Speed, in miles per hour (105 km/h), an adult human would have to run to stay on top of water as a basilisk does.

- Green basilisks escape predators by running across the water's surface.
- They stand on their hind legs, which rotate like windmills.
- Their long toes slap against the water, creating air pockets that keep the lizards above the water's surface.

Honeyguides Help Humans Hunt for Hives

It is rare for humans and wild animals to work together. But honeyguides and humans have worked together for millions of years. The honeyguide is a gray-and-white bird native to Africa. It teams up with humans to get beeswax, its favorite food.

Honeyguides lead humans to honey. This begins with a special call from a human. If a honeyguide is nearby, it will respond and lead the humans to a hollow tree. Inside is a bees' nest. The humans use smoke to calm the bees. Then they chop down the tree. The humans take the honey. The honeyguide gets the wax.

Honeyguide parents do not teach their young how to work with humans. Young honeyguides may learn from other adults. Or the behavior might be instinct. Honeyguides do not respond to just any human sound. They listen for the specific call.

Honeyguides respond to a specific call from humans.

Honeyguides and humans work together to find beehives.

A honeyguide communicates in turn with special chirps, tweets, and flutters. Scientists think this two-way relationship has existed for millions of years. It helps both species survive.

THINK ABOUT IT

Honeyguides work with humans toward a common goal. What other animals work with humans as honeyguides do?

8
Times, out of ten, that honeyguides led humans to a beehive in one scientific study.

- Honeyguides are gray-and-white birds native to Africa.
- They work and communicate with humans to find honey.
- The humans eat the honey and the honeyguides get the wax, their favorite food.

Hooded Seals Blow Big Balloons

The hooded seal lives in the coldest parts of the Atlantic Ocean and the Arctic Ocean. Its name comes from the saggy sac of skin on the front of the male's face. When the male feels challenged, the sac inflates. It forms a big, black balloon on the top of its head. The balloon flops down over the slope of the seal's nose.

A dark pink nose balloon inflates out of one of its nostrils.

Male seals use these unusual appendages to attract mates. Each spring, males inflate their hoods and nose balloons. They do this to impress female hooded seals.

Male hooded seals can inflate a saggy sac of skin on their foreheads.

THINK ABOUT IT

Male hooded seals use their hoods and nose balloons to attract potential mates. What other adaptations in the animal kingdom help different species find mates? Research to find your answer.

4
Age, in years, when male seals develop their hoods.

- The hooded seal is named for the male's large, black hood.
- Males can also inflate pink nose balloons.
- Males use their hoods and nose balloons to attract females and threaten rivals.

Sometimes several male seals compete for one female. Hooded seals are big animals. They are also aggressive. When two males compete, they inflate their hoods to look even more threatening. The smaller male usually backs down. But if the seals are the same size, they fight. They shove and chase each other until one gives up. The winner becomes the female's mate.

A male hooded seal with an inflated nose balloon

Hungry Dung Beetles Dine on Manure

Dung beetles are picky about the kind of feces they eat. More than 100 million years ago, they ate dinosaur droppings. They have evolved to feast on the feces of large mammals, such as elephants and cows. Dung beetles munch on tiny, nitrogen-rich particles found in mammal manure. Nitrogen helps build the beetles' muscles. Strong muscles are necessary for moving poop and digging tunnels.

Dung beetles might be tunnelers, rollers, or dwellers. Tunnelers make their homes in the soil underneath the dung. Rollers use their heads and antennae to roll feces into large balls. These balls can be as large as apples. The beetles bury the balls in their nests. Dwellers lay their eggs inside manure.

A dung beetle can eat more than its own weight of feces in a single day. That is a good thing. Dung beetles help the manure decompose, or break down. The broken-down manure adds nutrients to the soil. Other plants and animals need these nutrients to survive.

A roller dung beetle turns dung into a small ball.

Dweller dung beetles lay their eggs in animal feces.

1,141
Times its own weight the taurus scarab dung beetle, the strongest insect in the world, can pull.

- Dung beetles eat the feces of large mammals.
- There are three types of dung beetles: rollers, tunnelers, and dwellers.
- The dung beetle's diet helps manure break down faster, which is good for the environment.

THINK ABOUT IT

Some dung beetles lay their eggs inside a ball of dung. How might that help the larvae's survival?

8
Lampreys' Jawless Mouths Latch onto Prey

Vampires may be a myth. But bloodsucking animals are real. One of these animals is the lamprey. These eel-like fish glide through freshwater and salt water. They hunt for their next meal, usually a fish. Many animals eat fish. But few catch their prey as lampreys do.

Lampreys do not have jaws. But they do have mouths.

A lamprey's mouth is a circle covered in sharp, hooked teeth. An adult lamprey uses its teeth to latch onto its prey. The fish tries to break free. But the lamprey's powerful suction and hooked teeth make escape impossible.

Three big teeth in the center of the lamprey's mouth scrape away the fish's skin to reveal its blood

Bloodsucking lampreys do significant damage to their prey.

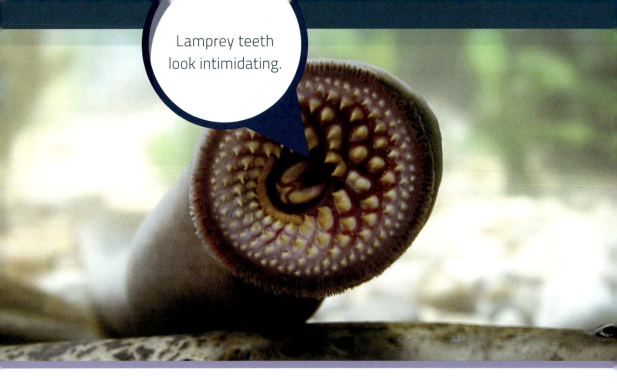

Lamprey teeth look intimidating.

vessels. The lamprey may feed for up to a whole day. Bloodsucking lampreys usually let their prey live. This is unusual for predators. The prey can rebuild its blood supply. They can become lamprey food again in the future.

LIVING FOSSILS

The lamprey has changed little in its 360 million years on Earth. It is known as a living fossil. Living fossils are animals that have very few, if any, living relatives. They look almost exactly the same as their ancient ancestors. Other living fossils include the aardvark and the platypus.

7
Years it takes the lamprey to reach adulthood.

- Bloodsucking lampreys feed on the blood of animals.
- Lampreys have mouths full of hooked teeth but no jaws.
- Bloodsucking lampreys do not kill their prey.

9

Narwhal Tusks Have Super Senses

The Arctic-dwelling narwhal is called the unicorn of the sea. This rare whale appears to have a long, spiraled horn attached to its face.

But the horn is actually a tooth. It is one of two in a narwhal's mouth. In females, both teeth grow to approximately 1 foot (30 cm) long.

But in males, the left tooth can grow up to 9 feet (2.7 m) long. It is so big that it actually pierces through the narwhal's lip.

Scientists are still trying to figure out the purpose of this gigantic tooth. Once, scientists believed narwhals used their tusks for defense.

A male narwhal in arctic waters near Nunavut, Canada

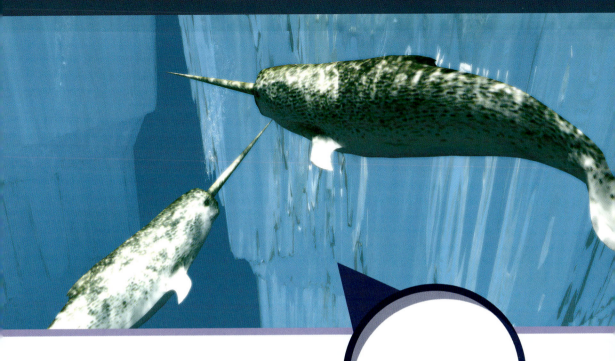

Male narwhal show off their long tusks.

Others thought the whales used their tusks to break through the ice.

Scientists' further research disproved those ideas. The tusk's structure indicates that it is a sensory organ. Unlike human teeth, it is not covered by a protective layer of enamel. Up to 10 million nerve endings are exposed at all times. This makes the narwhal's tusk very sensitive. Fighting and breaking through ice would be very painful. But the sensitivity is helpful in other ways. It helps the narwhal detect the temperature of the water. It can detect chemicals in the water. It may even allow narwhals to figure out if females are ready to mate.

50
Average number of years a narwhal lives.

- The male narwhal's long left tusk earned it the nickname the unicorn of the sea.
- Scientists once thought the tusk was for fighting or breaking ice.
- Scientists now know that the tusk is a giant sensory organ.

21

10

Opossums Stage Dramatic Death Scenes

Opossums are some of the fiercest-looking backyard creatures. They have more teeth than any land mammal in North America. An opossum's hiss keeps most predators away. But some predators are not frightened by the opossum's teeth and hiss. In these cases, the opossum takes the dramatic way out. It pretends to be dead.

Playing dead is not just for show. It is a survival tactic. When threatened, the opossum falls to the ground. Its tongue hangs out of its drooling mouth. Its beady eyes stare into the distance. It poops, releasing a stinky, oily green slime. The predator decides to find dinner elsewhere.

Opossums may look fierce, but they are more likely to play dead than attack.

> Playing dead helps opossums survive in a world full of predators.

The opossum may look and act dead. But it is actually very alert. It listens for more signs of trouble. Opossums are cautious animals. They might wait for hours before deciding it is safe to move. Once danger has passed, the opossum gets up as if nothing happened.

THE OPOSSUM'S SUPERPOWER

Opossums have existed for 65 million years. They have had lots of time to adapt for survival. One important adaptation is a protein in opossum blood. The protein stops most toxins from harming opossums.

50
Number of razor-sharp teeth in the opossum's mouth.

- Opossums play dead when predators threaten them.
- Though they are completely alert, they look and smell dead.
- Opossums can play dead for hours until they sense danger has passed.

11 Pangolins Protect Themselves with Armor

The pangolin looks like a cross between an anteater and an armadillo. It is native to southern Africa and Asia. Its tough scales are armor against predators. It uses its long, curved claws to dig burrows and destroy anthills.

The pangolin is a carnivore. But it does not have any teeth. It uses its tongue to snatch up ants, termites, and other small insects. Its tongue is attached to its long breastbone. Pangolins have terrible eyesight. They use their excellent senses of smell and hearing to find their prey.

Pangolins are the only mammals in the world covered in scales. The scales are made of keratin. This is the same protein found in human hair. Like hair, the scales are constantly growing. They protect the pangolin from predators such as lions, tigers, and leopards. When the pangolin feels threatened, it curls itself into a ball.

A pangolin searches for insects to eat.

70 million
Number of insects one pangolin can eat in a year.

- The pangolin is the only scaled mammal in the world.
- It eats ants, termites, and other small insects.
- When threatened by predators, it rolls up into a ball and lets out a stinky acid.

THE MOST HUNTED ANIMAL IN THE WORLD

The pangolin is the most hunted animal in the world. It is worth a lot of money in Asia. People use its scales for folk medicine and eat its meat. More than 100,000 pangolins are illegally captured and traded each year. If this trade continues, all eight pangolin species may soon be extinct.

Then, it emits a stinky acid. Only human hunters are willing to overcome this smelly defense.

Rolled up, a pangolin looks more like a pine cone than an animal.

Sneaky Octopuses Outsmart Predators

Octopuses are the smartest invertebrates in the world. They live in warm ocean waters around the globe. An octopus's arms carry more than two-thirds of its neurons. Neurons are cells that transmit information to the brain. Having lots of neurons in its arms helps the octopus. The arms can perform tasks without instructions from the brain. This makes the octopus a good multitasker. It can keep an eye out for trouble while its arms open clamshells.

Octopuses are masters of camouflage. Special skin cells called chromatophores allow them to change color instantly. Octopuses change colors to show emotion. They might change to scare away nosy fish. They can also change color to blend in with their surroundings. This makes them nearly invisible to hungry predators, such as sharks and dolphins.

If spotted, the octopus shoots a dark, inky substance into the water. This temporarily blinds the predator. It also confuses the predator's senses of taste and smell. A predator

Camouflage makes it difficult to spot the octopus in this photograph.

INKY THE ESCAPE ARTIST

An octopus made headlines in early 2016. Inky the octopus made a daring escape from the National Aquarium in New Zealand. One evening, someone did not fully close the lid on Inky's tank. Inky pushed the lid aside and climbed out of the tank. He crawled across the floor to a drainpipe. His boneless body allowed him to fit through the narrow pipe. The pipe led him straight to the ocean. Scientists discovered the wet trail Inky left behind. But they never found Inky.

that gets too close may get a bite from the octopus's sharp beak. If a predator bites, the octopus has a last resort. It can detach an arm from its body. The arm will grow back later, as good as new.

3
Number of hearts an octopus has.

- Octopuses are the most intelligent invertebrates in the world.
- They use camouflage to hide from predators and prey.
- When threatened, octopuses spray ink, bite, and even detach their arms.

An octopus shoots ink at a diver trying to catch it.

Fact Sheet

- There are four types of camouflage: concealing, disrupting, disguising, and mimicry. Concealing camouflage is coloring that allows an animal to blend in with its surroundings. Disrupting camouflage is a pattern on an animal's skin that makes it hard for a predator to spot the animal. Disguising camouflage is a texture that helps an animal blend in with its surroundings. Mimicry involves an animal looking or acting like another animal.

- Everyone has to eat. Many animals have adapted their bodies and their diets to best meet their needs for survival. Giraffes have evolved long necks to reach leaves high in the trees. Their long tongues grab leaves and shoots and pull the vegetation into their mouths. The blue whale is the largest animal in the world. It eats tiny animals called krill. It uses fringed pieces of fingernail-like material called baleen to strain krill from the water.

- The survival of a species depends on the ability of adults to reproduce. Some adaptations increase the likelihood of future generations. Male bowerbirds build elaborate structures out of sticks and blue objects. Then, they dance for the female birds they are trying to woo.

- Many animals have developed astonishing ways of protecting themselves from predators. Vultures vomit when they feel threatened. Texas horned lizards shoot blood out of their eyes. A South American termite species explodes when their nest is attacked.

Glossary

adaptations
Changes animals make to increase their chances of survival.

adapted
Changed to increase chances of survival.

appendages
Parts of an animal attached to the main body, such as arms or legs.

bioluminescence
A chemical process that makes some animals glow.

brood parasites
Animals who rely on other species to raise their young.

evolved
Changed over time.

feces
The bodily waste of an animal.

instinct
A natural behavior.

invertebrates
Animals that do not have backbones.

nutrients
Substances that support life.

predator
An animal that kills and eats another animal.

prey
An animal that is killed and eaten by another animal.

vessels
Veins, arteries, and capillaries that carry blood in the body.

For More Information

Books

Arnold, Caroline. *Living Fossils: Clues to the Past.* Watertown, MA: Charlesbridge, 2016.

Burnie, David. *The Animal Book: A Visual Encyclopedia of Life on Earth.* New York: DK, 2013.

Johnson, Rebecca L. *Journey into the Deep: Discovering New Ocean Creatures.* Minneapolis: Millbrook, 2011.

Spelman, Lucy. *Animal Encyclopedia: 2,500 Animals with Photos, Maps, and More!* Washington, DC: National Geographic, 2012.

Visit 12StoryLibrary.com

Scan the code or use your school's login at 12StoryLibrary.com for recent updates about this topic and a full digital version of this book. Enjoy free access to:

- Digital ebook
- Breaking news updates
- Live content feeds
- Videos, interactive maps, and graphics
- Additional web resources

Note to educators: Visit 12StoryLibrary.com/register to sign up for free premium website access. Enjoy live content plus a full digital version of every 12-Story Library book you own for every student at your school.

Index

Africa, 12, 24
anglerfish, 4–5
Arctic Ocean, 14, 20
Asia, 24, 25
Atlantic Ocean, 14
atlas moths, 6–7

bioluminescence, 4

camouflage, 11, 26
Central America, 10
chromatophores, 26
cuckoos, 8–9

dung beetles, 16–17

green basilisks, 10–11

honeyguides, 12–13
hooded seals, 14–15

lampreys, 18–19

mimics, 6, 7

narwhals, 20–21
National Aquarium, 27
North America, 22

octopuses, 26–27
opossums, 22–23

pangolins, 24–25

Southeast Asia, 6

About the Author
Kristin Marciniak writes the books she wishes she had read in school. A graduate of the University of Missouri School of Journalism, she lives with her family in Overland Park, Kansas. She loves learning about the wildlife in her backyard, including opossums, turtles, and bats.

READ MORE FROM 12-STORY LIBRARY

Every 12-Story Library book is available in many formats. For more information, visit 12StoryLibrary.com.